AR 4.6/0.5

HOW CAN I EXPERIMENT WITH ... ?

ATOMS

Cindy Devine Dalton

Cindy Devine Dalton graduated from Ball State University, Indiana, with a Bachelor of Science degree in Health Science. For several years she taught medical science in grades 9-12.

Teresa and Ed Sikora

Teresa Sikora teaches 4th grade math and science. She graduated with a Bachelor of Science in Elementary Education and recently attained National Certification for Middle Childhood Generalist. She is married with two children. Ed Sikora is an Aerospace Engineer, working on the Space Shuttle Main Engines. He earned a Bachelors of Science degree in Aerospace Engineering from the University of Florida and a Masters Degree in Computer Science from the Florida Institute of Technology.

Rourke Publishing LLC
Vero Beach, Florida 32964

PROJECT EDITORS
Teresa and Ed Sikora

PHOTO CREDITS
Corel
Gibbons Photography
PhotoDisc
Walt Burkett, Photographer

ILLUSTRATIONS
Kathleen Carreiro

EDITORIAL SERVICES
Pamela Schroeder

Library of Congress Cataloging-in-Publication Data

Dalton, Cindy Devine, 1964–
    Atoms / Cindy Devine Dalton.
        p. cm. — (How can I experiment with?)
    Includes bibliographical references and index
    ISBN 1-58952-010-6
    1. Matter—Constituion—Juvenile literature. 2. Atoms—Experiments—Juvenile literature. I.Title

QC173.16.D35 2001
539.7—dc21   00-068353

Printed in the USA

Atoms: Tiny particles made up of protons, neutrons, and electrons.

Quote:

"All the world is a laboratory to the inquiring mind."

-Martin H. Fischer

Table of Contents

What Is Matter?

Everything on Earth is made up of **matter**. Matter takes up space. Matter stays the way it is until something forces it to change. Think of a swimming pool. The water stays still until wind, rain, or even a person makes waves. The way the water changes depends on the force that makes it change. Rain hitting the pool affects the water differently than you jumping in.

Everything around you is made up of matter. The air, animals, and even people are made of matter.

Breaking Matter Down

Now that we know what matter is, let's think of matter in its smallest form. You even need a special microscope to see it. Matter is made up of tiny particles called **atoms.** There are three parts to an atom.

positive (+) **proton**

negative (—) **electron**

neutral (=) **neutron**

Atoms are too tiny to be seen without a special microscope.

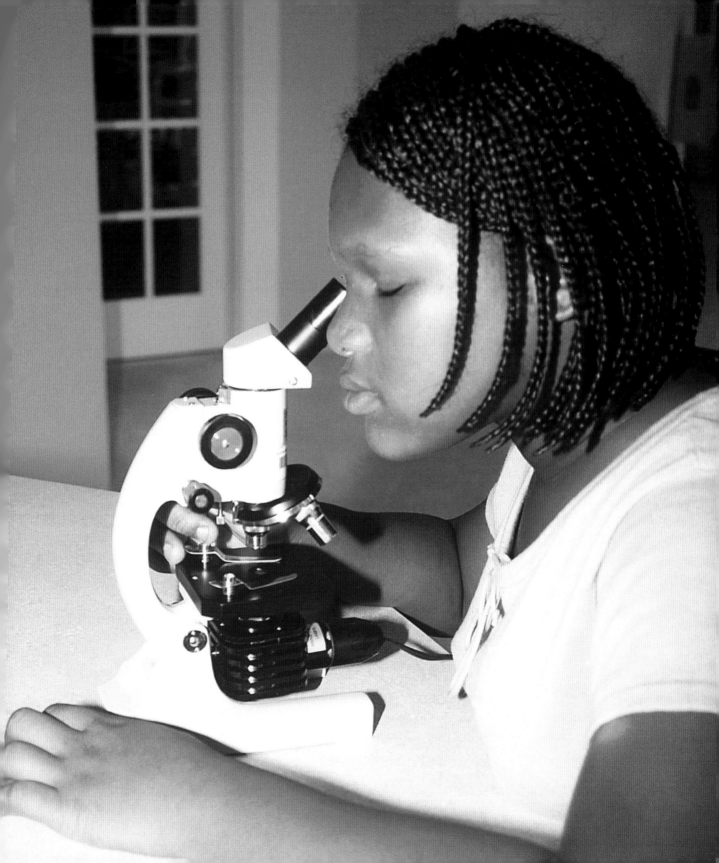

Atoms, Atoms, Everywhere

Everything is made up of atoms. You are made up of atoms. Your pets are made up of atoms. Even your favorite toys are made up of atoms.

There are only about 90 different types of atoms in nature. **Scientists** have made 25 more.

This chart shows the different amounts of protons and electrons in each element. Each one is a different substance.

1 H																	2 He
3 Li	4 Be											5 B	6 C	7 N	8 O	9 F	10 Ne
11 Na	12 Mg											13 Al	14 Si	15 P	16 S	17 Cl	18 Ar
19 K	20 Ca	21 Sc	22 Ti	23 V	24 Cr	25 Mn	26 Fe	27 Co	28 Ni	29 Cu	30 Zn	31 Ga	32 Ge	33 As	34 Se	35 Br	36 Kr
37 Rb	38 Sr	39 Y	40 Zr	41 Nb	42 Mo	43 Tc	44 Ru	45 Rh	46 Pd	47 Ag	48 Cd	49 In	50 Sn	51 Sb	52 Te	53 I	54 Xe
55 Cs	56 Ba		72 Hf	73 Ta	74 W	75 Re	76 Os	77 Ir	78 Pt	79 Au	80 Hg	81 Ti	82 Pb	83 Bi	84 Po	85 At	86 Rn
87 Fr	88 Ra		104 Rf	105 Db	106 Sg	107 Bh	108 Hs	109 Mt	110 Uun	111 Uuu	112 Uub	113 Uut	114 Uuq	115 Uup	116 Uuh	117 Uus	118 Uuo

57 La	58 Ce	59 Pr	60 Nd	61 Pm	62 Sm	63 Eu	64 Gd	65 Tb	66 Dy	67 Ho	68 Er	69 Tm	70 Yb	71 Lu
89 Ac	90 Th	91 Pa	92 U	93 Np	94 Pu	95 Am	96 Cm	97 Bk	98 Cf	99 Es	100 Fm	101 Md	102 No	103 Lr

John Dalton

Scientists started studying atoms hundreds of years ago. John Dalton, a scientist, developed the first useful **theory** about atoms in 1803. He was the first to say that everything in the world is made of atoms. He believed atoms were all alike. By putting atoms together in different ways, you can make all the things on Earth.

We've learned more since John Dalton wrote his first theory of atoms.

This is what scientists in the early 1900s thought an atom looked like.

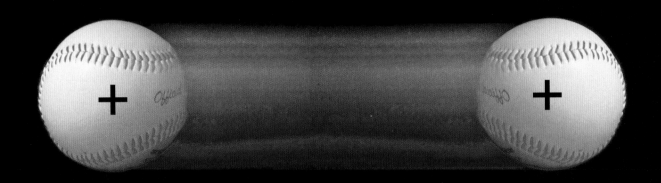

If your baseballs were protons they would repel each other.

If one of your baseballs was a proton and the other was an electron they would attract each other.

What Would You Think?

Imagine you threw two baseballs straight up into the air with the same force and called them **protons**. Now, instead of coming back down to you they went sideways, away from each other. What would cause the baseballs to do that?

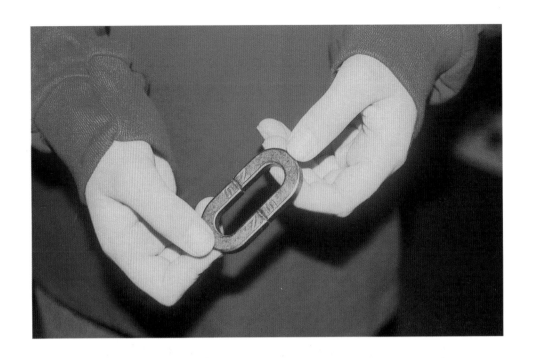

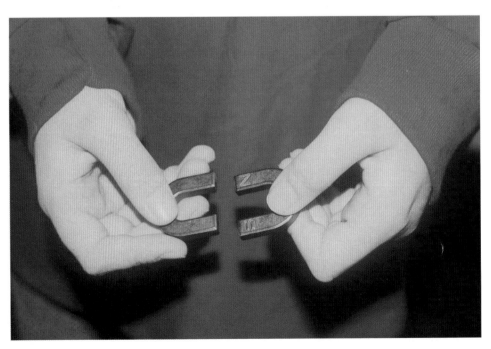

16 *Atoms are attracted to a charge that is different than their own. You can use magnets to experiment.*

Opposites Attract!

One law of atoms is that parts that are alike **repel** each other, or go in opposite directions. Remember the two baseballs from the first chapter? We will call them protons. Protons do not attract each other. They are both positive. That is why the baseballs went away from each other. Electrons do not attract each other either because they are all negative.

However, protons do attract electrons and electrons do attract protons. Positive and negative are opposites. Another law of atoms is that opposites attract. If your baseballs had been different, one a proton and one an electron, they would have stayed together.

What Does an Atom Look Like?

Atoms are like eggs. The yoke in an atom is called the **nucleus.** Inside the nucleus are the protons and neutrons. Going around the nucleus are the electrons. Atoms are very, very small. However, their charge can be very strong.

There are some parts of you that make you very special. There are parts of an atom that make it special, too. That part is the number of protons and electrons it has. Each kind of atom has a different number of protons and electrons. If you add one or take one away, it becomes a new type of atom.

The blue area in this diagram is the nucleus. Going around the nucleus in the yellow area are the electrons.

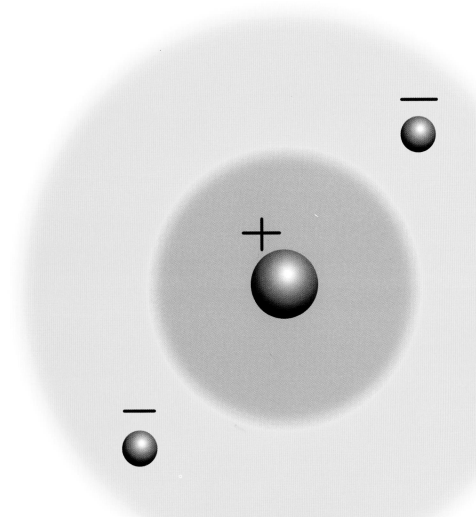

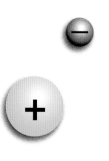

HYDROGEN

1-Proton
1-Electron
0-Neutron

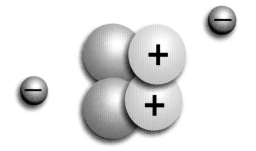

HELIUM

2-Protons
2-Electrons
2-Neutrons

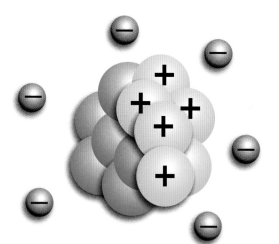

CARBON

6-Protons
6-Electrons
6-Neutrons

Peeling Away the Layers

Have you ever peeled an onion? An atom is made like an onion. It has layers. The electrons go around the nucleus in layers. Electrons can jump from layer to layer if a force affects them. Every atom has a different number of electrons. This number tells us the charge of the atom.

The number of protons, electrons, and neutrons in an atom helps us know what type of atom it is.

PROTON

—Protons carry a positive charge.

NEUTRON

—Neutrons carry no charge.

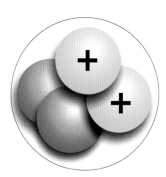

PROTONS and **NEUTRONS** join
together to form the **NUCLEUS**—
the central part of the atom

This is called the *Quantum Theory*. Niehls Bohr developed this theory in 1913. Later, in 1932, scientists discovered that neutrons were inside the nucleus with protons. Neutrons are neutral. They have no charge. They were more difficult for scientists to find.

In 1932 scientists discovered that neutrons were inside the nucleus with protons.

Chemical Reactions

Chemical reactions happen all day long all around us. A chemical reaction is when one substance is changed into another substance. The change occurs because groups of atoms get connected, or break apart. Starting a fire is a chemical reaction. Digesting food is another one. Whenever two atoms are either combined or separated, you have a chemical reaction, and a new substance is created. Some of the time chemical reactions give off heat.

Food causes many chemical reactions in your body. Ice cream cools the tiny atoms of your mouth.

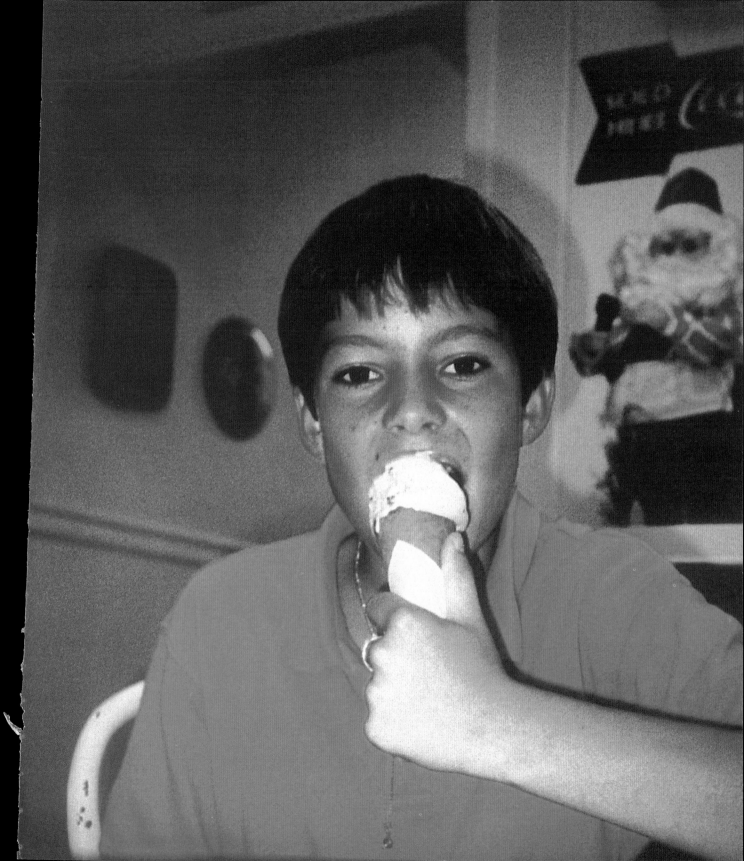

The Atomic Bomb

The atomic bomb was produced at Los Alamos, New Mexico. It was made when scientists combined substances that create a chain reaction. The chain reaction causes a huge amount of heat energy.

The first atomic bomb was dropped on the city of Hiroshima, Japan on August 6, 1945. The bomb killed thousands of people. The substance that was created because of the chemical reaction was also deadly. The bomb not only killed people, but also poisoned the land, animals, and air.

The atomic bomb is one of the most destructive forces known to man.

27

"The Atoms Family"

You can sing these words to the tune of the "Addams Family" song. However, the words are about the Atoms Family.

They are so small. They're round like a ball.
They're in the air. They're everywhere.
You can't see 'em at all.

They're tiny and they're teeny. They're much
smaller than a beanie. They never can be
seen-ie, the Atoms Family.

They are so small. They're round like a ball.
They're in the air. They're everywhere.
You can't see 'em at all.

They make up all the gases, and liquid like
molasses, and all the solid masses,
the Atoms Family.

Trivia

Every person has at least 40,000 atoms changing into different atoms in his or her body each second.

Each human body has about 60 trillion cells. Each body cell contains an average of 90 trillion atoms, or more than 10,000 times as many molecules as the Milky Way has stars.

The number of atoms in 1 pound (373 grams) of iron is almost 5 trillion billion.

The pressure at the center of the sun (700 million tons per square inch) is enough to smash atoms.

Glossary

atom (AT em) — a tiny particle made up of protons, neutrons, and electrons

electron (ih LEC tron) — a part of an atom that has a negative charge and travels around its nucleus

matter (MAT er) — what everything is made of

nucleus (NOO klee es) — the central part of an atom

neutron (NOO tron) — the part of an atom that has no charge

proton (PROH ton) — the part of an atom that is inside the nucleus and has a positive charge

repel (re PEL) — to refuse or push back

scientist (SY en tist) — person who studies science

theory (THEE eh ree) — an idea about how the world works

Websites to visit

*www.spartechsoftware.com/reeko/Experiments/Exp
AttractingTape.htm*
*www.antoine.frostburg.edu/chen/senese/101/atoms/
dalton/shtml*
www.encarta.msn.com

Index